SAE EDGE™
RESEARCH REPORT

Unsettled Issues on Human-Robot Collaboration and Automation in Aerospace Manufacturing

Professor Philip Webb, PhD
Cranfield University
Sarah Fletcher, PhD
Cranfield University

EDGE DEVELOPMENT TEAM

Andrew Armstrong, *Fanuc Robotics*
Marco Chacin, PhD, *Airbus*
Elena Dominguez, *Pilz*
Joshua Downs, *Boeing*
Leo Muijs, *GKN Fokker*
Andrew Portsmore, *GKN Aerospace*

SAE INTERNATIONAL®
Warrendale, Pennsylvania, USA

INTERNATIONAL®

About the Publisher

SAE International® is a global association of more than 128,000 engineers and related technical experts in the aerospace, automotive and commercial-vehicle industries. Our core competencies are life-long learning and voluntary consensus standards development. Visit sae.org

SAE EDGE™ Research Report Disclaimer

SAE EDGE™ Research Reports focus on topics that are dynamic, in which knowledge is incomplete, and which have yet to be standardized. They represent the collective wisdom of a group of experts and serve as a practical guide to the reader in understanding unsettled subject matter. They are not meant to provide a recommended practice or protocol. The experts have assembled as a community of practitioners to contribute and collectivize their thoughts and points of view; these are not the positions of the institutions or businesses with which they are affiliated, nor is one contributor's perspective advanced over others. SAE EDGE™ Research Reports are the property of SAE International and SAE alone is responsible for their content.

About This Publication

SAE EDGE™ Research Reports provide state-of-the-art and state-of-industry examinations of the most significant topics in mobility engineering. SAE EDGE™ contributors are experts from research, academia, and industry who have come together to explore and define the most critical advancements, challenges, and future direction in areas such as vehicle automation, unmanned aircraft, cybersecurity, advanced propulsion, advanced manufacturing, Internet of Things, and connectivity.

Related Resources

SAE MOBILUS® Advanced Manufacturing Knowledge Hub
https://saemobilus.sae.org/advanced-manufacturing/

SAE MOBILUS® Automated & Connected Knowledge Hub
https://saemobilus.sae.org/automated-connected/

SAE Team

Frank Menchaca, Chief Growth Officer
Michael Thompson, Director of Standards, Information and Research Publications
Monica Nogueira, Director of Content Acquisition
Beth Ellen Dibeler, Product Manager
William Kucinski, Managing Technical Editor

Printed in USA

EPR2020024
ISSN 2640-3536
e-ISSN 2640-3544
ISBN 978-1-4686-0254-8

To purchase bulk quantities, please contact: SAE Customer Service

E-mail: CustomerService@sae.org
Phone: 877-606-7323 (*inside USA and Canada*)
+1-724-776-4970 (*outside USA*)
Fax: +1-724-776-0790

https://www.sae.org/publications/edge-research-reports

About the Editors

Cranfield University

Professor Phil Webb is the head of the Centre for Structures, Assembly and Intelligent Automation at Cranfield University, where he holds the Royal Academy of Engineering and Airbus Chair.

He has over 25 years of experience in research into advanced manufacturing automation and robotics, and has published widely on the subject. He specializes in the application of automation in the aerospace industry and works with most of the major aerospace companies including Airbus, BAE Systems, Bombardier, and Rolls-Royce. A core element of this work is the development of human-robot collaborative systems, particularly using large, high-payload robots. Professor Webb has published in this area and has developed numerous production demonstrators as well. His role as an Airbus Professor sits between academia and industry giving him the opportunity to bring emerging engineering research across into the manufacturing domain.

He is a Fellow of the Institution of Engineering and Technology, a member of the SAE International Automated Assembly and Fastening Committee, and a member the Council of the British Automation and Robotics Association. He has also served as the UK representative to the International Federation for Robotics. He sits on the ATI Future Structures Specialist Advisory Group and is the Editor in Chief of the *Aircraft Engineering and Aerospace Technology Journal*. He has served as an organizer and contributor to SAE's AeroFast and AeroTech conferences for many years.

Cranfield University

Dr. Sarah Fletcher is a Senior Research Fellow at Cranfield University, where she leads the Industrial Psychology and Human Factors Group.

Sarah has been conducting human-systems interaction research for 20 years, primarily within the aerospace manufacturing industry. She specializes in cognitive and performance analysis in industrial automation and human-robot collaboration contexts, and the application of mixed methods for human analysis including sensor-based tools. She has led Human Factors research on many UK and EU projects and teaches a range of topics related to the psychology and ethics of robotics and artificial intelligence. She sits on a number of standards committees, including ISO/TC 299/WG3 Industrial safety, ISO/TC 159/SC4 Ergonomics of human-system interaction, and AMT/10/1 Ethics for Robots and Autonomous Systems.

contents

Atosan/Shutterstock.com

Unsettled Issues on Human-Robot Collaboration and Automation in Aerospace Manufacturing

Abstract

Is human-robot collaboration the future for aerospace manufacturing? It is widely acknowledged that the application of robotics and automation in aerospace manufacturing is significantly lower than might be expected given the size and value of the industry and its technologically advanced products. When deployed, it tends to be either in relatively simple activities, such as machine loading and handling, or in the form of large bespoke systems for processes such as fastening. The reason for this is multifactorial but includes product variability, size, design philosophy, and relatively low volumes. Also, there is occasional reticence due to a history of past false starts, which increases the perceived risk associated with the introduction of new technologies.

Current thinking suggests that the emerging technology of human-robot collaboration provides an ideal solution, combining the flexibility and skill of human operators with the precision, repeatability, and reliability of robots. However, this is a topic that tends to generate emotional and emotive reactions within the industry ranging in extremes from a "brave new and inevitable future for aircraft manufacturing" to assembly to workers living in fear of a robot invasion and lost jobs. In this SAE EDGE Research Report, we seek to build a comprehensive picture of the current state-of-the-art of human-robot applications and identify key issues that unlock the technology's potential. We have sought the views of recognized thought leaders to understand and deconstruct the myths and realities of human-robot collaboration, and how it could eventually have the impact envisaged by many.

NOTE: SAE EDGE™ Research Reports are intended to identify and illuminate key issues in emerging, but still unsettled, technologies of interest to the mobility industry. The goal of SAE EDGE™ Research Reports is to stimulate discussion and work in the hope of promoting and speeding resolution of identified issues. SAE EDGE™ Research Reports are not intended to resolve the challenges they identify or close any topic to further scrutiny.

PROFESSOR PHILIP WEBB, PhD
Cranfield University

SARAH FLETCHER, PhD
Cranfield University

Edge Development Team

Andrew Armstrong, *Fanuc Robotics*
Marco Chacin, PhD, *Airbus*
Elena Dominguez, *Pilz*
Joshua Downs, *Boeing*
Leo Muijs, *GKN Fokker*
Andrew Portsmore, *GKN Aerospace*

ISSN 2640-3536

Introduction

State of the Industry

Aerospace manufacturing is generally characterized by the size and complexity of the product to be manufactured—in other words, their relatively low volume and high precision. However, this is a generalization as many individual parts and subassemblies are involved for which the volumes can become quite high though still not approaching, for example, automotive or white goods manufacture. Indeed, there has been a significant drive over the last few years to map and transfer best practice manufacturing technologies from automotive to aerospace. But, in reality, this has been limited except for areas that do not directly involve the product, such as lean manufacturing principles. The entire design for manufacture and assembly philosophy is different and more cost-focused against the strict design and performance constraints on aerostructures (Figure 1). The most notable difference is the limited use of automation, though in the automotive industry, the automation focus is often on "body in white" and engine assembly, while many of the finishing and interior fit out processes are still largely manual.

In aerospace, automation has a patchy and somewhat problematic history with many failed programs, the reasons for which are multiple and varied. For instance, the size and delivered dimensional quality variability of the parts against the expected nominal dimensions of the structure is a significant barrier that relies on human skill and knowledge to correct during assembly. Another key issue is the failure to understand the complexity of many processes, and how much they rely on operator skill; this is particularly true in situations such as sealant application (which is commonly automated in other industries but has only been automated in limited applications and with limited success in aerospace). Similarly, reallocating process tasks is often precluded or unsuccessful because cognitive human skills, particularly those involving perception and decision-making, are not identified and understood well enough to enable accurate evaluation of which should remain with the operator versus which should be transferred to the robot. Not fully understanding human aspects also means that impacts on safety and well-being are unclear. This constrains risk assessments and diminishes an

FIGURE 1. **Conventional assembly facility.**

Skycolors/Shitterstock.com

organization's confidence that a new system will be acceptable to their workforce. When automation does occur, it is usually in the form of large "monument" machines that are capital intensive and somewhat inflexible. However, robots are used more widely in the supply chain applications such as machine loading, but these are not industry specific and are relatively simple and well established.

There has been significant interest and research in recent years into human-robot collaboration [1]. However, much of it has been focused on the service and assistive technology sectors. While there are some lessons that can be transferred, we will focus on the current representative state of the art in aerospace and manufacturing. Recent examples include the Validation of Advanced, Collaborative Robotics for Industrial Applications, or "VALERI," project funded by the European Commission, which demonstrated a mobile collaborative robot specifically designed for aircraft assembly processes, such as sealant application [2]. Additionally, work within Fokker has demonstrated the application of collaborative robotics to a number of tasks including drilling and sealing. Of particular interest is an application which demonstrates how an existing hand tool—an automated drilling unit—can be picked up and manipulated by a mobile robot to perform a drilling operation on an aircraft. This has the potential for common tooling between human operator and robot, thus significantly reducing the need for costly end effectors and allowing resources to be shared [3]. The Boeing Company in the United States has also implemented a number of individual collaborative robots within its manufacturing facilities (e.g., the Future Automated Upright Build system installed to assemble the fore and aft sections of the 777 fuselage). The collaborative robots work alongside the human operators helping to perform both drilling and installation processes [4]. However, Boeing announced the partial decommissioning of this collaborative installation in 2019 due to technical problems associated with the robotic element of the system [5].

Airbus also has a number of research and development programs related to collaborative robots, many of which relate to the implementation of humanoid robots, but the majority are more conventional in their configurations (Figure 2) [6].

This is by no means an exhaustive list of applications, but all of these applications share the common goal of seeking to reduce or augment the use of human labor with varying degrees of success. However, given the size of the industry, these types of approaches still represent a minuscule uptake of collaborative technology (Figure 3).

Unsettled Issues on Human-Robot Collaboration and Automation in Aerospace Manufacturing

The emerging technologies associated with human-robot interaction have often been suggested as a possible solution to

FIGURE 2. **Collaborative robot integration with existing manufacturing tooling.**

Ekaterina Kupeeva/Shutterstock.com

FIGURE 3. How close will interaction become?

some of the inherent problems in aerospace manufacturing. Indeed, this is also the case for many applications across the wider manufacturing industry that have so far been resistant to automation. However, uptake has been fraught with problems and false starts. Within this SAE EDGE Research Report, we will attempt to tease out and explore what are the important unsettled issues that are limiting its impact within the industry. The issues identified are as follows:

- **Defining collaborative robotics**

  There are many definitions and interpretations of what collaborative robotics is and how they vary depending on different applications and local perceptions. The term "cobot" or "cobotics" is often used to describe a subset of robots that are collaborative, but this is actually meaningless as all robots are capable of collaboration depending on their configuration and application. We need to establish a taxonomy and way of properly assessing collaboration.
- **Defining what makes a robot collaborative**

  Regardless of definition, all robots have collaborative potential. Some have their structures and/or control systems optimized to make them easier to deploy in collaborative roles. We need to shift toward considering the technology as an application rather than a type.
- **Effective application and task division between humans and robots**

  To fully exploit human-robot collaboration, we need to have a clear understanding of how we can divide the proposed task or process between the two elements. Simply allocating the most basic tasks to the robot and the complex ones to the human is not necessarily the best approach. Also, we need to consider carefully whether human-robot collaboration is actually the most appropriate course. We need to establish a clear methodology to first validate if a task is suitable for collaborative working and, second, for the optimal selection and division of tasks between humans and robots.
- **Building an effective and credible safety case for the implementation of collaborative robots**

  The safety of collaborative robots is often questioned, but the reason is not always based on rational analysis. While safety when working with robots must always be paramount and is absolutely critical, we need to ensure that decisions are informed by proper rigorous and legally justifiable criteria. In reality, decisions are often based on partial understanding of their capability and the legislation as well as historical experience and anecdotal concerns. Available guidance and safety standards focus on the prevention of physical injury and include little or no information on best practices for integration to optimize workforce safety behaviors.
- **Understanding the human aspects of human-robot collaboration**

  The fundamental aim of collaborative robotics is to combine and utilize the unique strengths of both humans and robots to optimize performance. In almost all cases, the human role is to supply cognitive skills and flexibility (i.e., decision-making, monitoring, remedial interventions, etc.). However, despite this fundamental purpose, little is actually understood regarding these crucial cognitive reactions or how we can optimize them. For collaborative robotics to be applied successfully, we need a much better understanding of design and implementation requirements, and the specific factors that will optimize workforce acceptance and adoption.

Defining Collaborative Robotics

There are many existing robot cells that have some degree of interaction with human operators. Currently, most applications involve "coexistence" where the operator works very close to a robot but the robot has its own workstation. These setups may require the operator to, for example, install parts into a jig prior to welding. The operator usually accesses the task through a hatch or other aperture, which minimizes the risk of physical contact. There are significant variations in this arrangement but, in general, the operator does not have free access to the cell even when the manipulator is stationary. However, to achieve the true potential of collaborative operation we need to achieve more subtle interactions, which in turn require much more complex definitions. Although collaborative robotics always involves the co-location of a human and a robot, the term is often used as if it is a single entity. Yet, it is not just one thing. Collaborative robotics includes a wide array of systems with different characteristics, particularly regarding types and degrees of human-robot interaction. It is worth us considering examples of these differences because they are—to some extent—the underlying cause for confusion and unsettled issues. It is clear that within the aerospace industry (and beyond) there is no unified view or taxonomy that defines collaborative robotics. Discussions with industry experts have shown that while descriptions vary, there are some unifying characteristics and common emerging themes:

- Where a human and a robot work together, such that their effective working zones overlap (supporting a co-worker)
- Where a robot performs the same task as a human and they are working next to each other (potentially replacing a co-worker)
- Where a robot performs a specific task in between humans and is seen as part of a team (one of the co-workers)
- Where there is some interaction between the human and robot whether through touch, or commands

These are all valid descriptions and share common ideas of mutual workspace and tasks as well as interaction. The task is not necessarily an identifying characteristic and may vary significantly in its complexity. We will cover the task-dependent aspects in a later section. If we consider now the literature and current academic thought, we then find a similar but more detailed taxonomy; but again, different researchers identify specific nuances, which sometimes contradict one another.

Types of Interaction

Collaboration is one of three broad categories of human-robot interaction, which reflect increasing levels of physical contact and proximity [7]:

- **Supportive**: The robot assists the human by supplying equipment or information but does not directly take part in task performance; interactions are infrequent and transactional.
- **Collaborative**: The human and robot both work on different parts of the task, utilizing their respective skills; interactions are transactional but more continuous/frequent.
- **Cooperative**: The robot and human work together with shared responsibility for task performance; interaction is continuous, with direct or indirect physical contact.

Thus, although we now commonly refer to "collaborative" robotics, the more the human and robot become integrated to share task responsibilities and interactions the more the interactions become "cooperative."

Degrees of Interaction

All collaborative systems involve co-location of the human and robot, but the type and degree of interaction in each case can be very different due to the extensive range of system characteristic variables. Yanco and Drury's extended taxonomy of human-robot interaction demonstrates this point with eleven categories of different and variable characteristics [8]:

- Task type
- Task criticality
- Robot morphology (anthropomorphic, zoomorphic, or functional)
- Ratio of humans to robots
- Composition of robot teams
- Level of shared interaction among teams
- Interaction roles
- Type of human-robot physical proximity
- Decision support for operators
- Time/space taxonomy
- Autonomy level/amount of intervention

The range of possible combinations of system characteristics makes it difficult to apply general principles. Every technical system is potentially unique—even without the addition of human idiosyncrasies—and as technologies continue to advance this variability will only continue to grow.

Another broader approach is given below, which focuses on more generic aspects of the task and robot and also includes what might be considered as conventional robot systems:

- **Interruption**: This is simplest and most common in which an operator is able to stop or pause the operation of a robot (e.g., an emergency stop).
- **Intervention:** Similar to interruption but the operator will normally take a specific action and then restart the process.
- **Interaction:** This is the most complex process where the operator actually interacts with the robot and assists with the tasks.

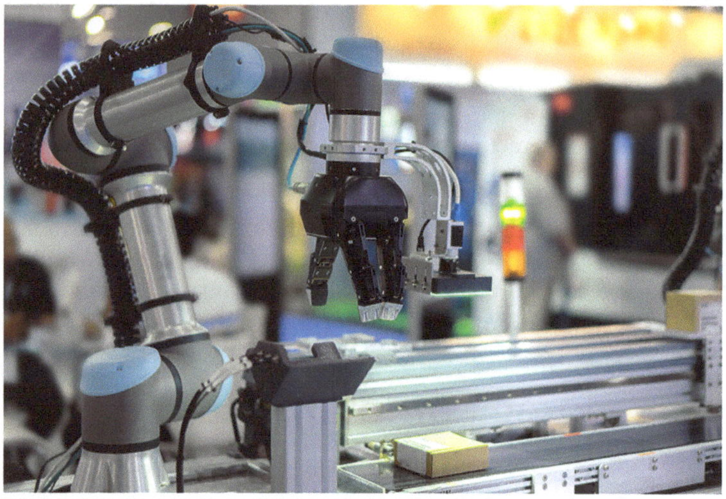

FIGURE 4. **Low payload collaborative robot.**

Both taxonomies are valued, though the first gives a much more specific and indicative description of the interaction despite its limited breadth of applicability and translatability as compared to conventional robot systems. How current thinking is moving away from treating collaborative robots as a special case but rather a subset of industrial robotics will be discussed later. This suggests that more general criteria are required.

Passive Interaction

Another type of interaction which we will only briefly cover here is what could be termed passive interaction. There are many existing and well-established robot cells that have some degree of interaction with human operators. These tend to use the human operator for tasks like installing parts into a jig prior to welding. However, the difference lies in the degree of separation; the main body of the cell is physically isolated from the operator by some type of barrier, either electrotechnical or physical (caging). The operator usually accesses the task through a hatch or other aperture, which minimizes the risk of any physical contact. There are significant variations in this arrangement but, in general, the operator does not have free access to the cell even when the manipulator is stationary.

Safe Shared Workspaces

A further aspect of collaboration is that it—by necessity—involves the safe sharing of a common workspace. Either the robot and the human operator are active at the *same* time in a common workspace jointly performing a task or the robot and the human operator are active at *different* times in a common workspace jointly performing a common task. In both cases, it is normally assumed that the robot will be reaming in automatic mode.

Recommendations

There is clearly a requirement for a unified taxonomy that can be used to define collaboration across different domains. This needs to link to the wider considerations such as standards and robot functionality, and applications to ensure that there is clear understanding of what a collaborative robot system is and what the true level and complexity of interaction are (Figure 4).

Defining What Makes a Robot Collaborative

All robots are potentially collaborative depending on the application and the way in which they are configured. A subset of robots has been designed that incorporate either extra sensing or control or mechanical features that are intended to optimize them for collaborative operation. These are sometimes referred to as cobots but, as previously discussed, this terminology will not be used in this report. The second category are applications which use advanced external sensing to provide a flexible "electronic" barrier between the human operator and robot. As identified in the previous section, we have multiple, different taxonomies and methodologies which—while all describing some form of collaborative hardware system—vary significantly in interpretation. What is becoming clear is that there are a number of requirements that are not currently being satisfied by the marketplace; some of which include the following:

- Greater autonomy of collaborative robots in order to improve the business case (less operator involvement)
- Specific end effectors for collaborative robots (safety aspects)

TABLE 1. Indicative collaborative robots.

Manufacturer	Name	Axes	Max payload (kg)
FANUC	CR series	6	4-35
Universal Robots	UR16,10,5,3	6	16/10/5/3
KUKA	LBR iiwa	7	7-14
ABB	YuMi (twin arm configuration)	14	0.5
Comau	AURA	6	170
Yaskawa	MOTOMAN HC10	6	10
Mitsubishi	MELFA ASSISTA	6	5

- Stronger collaborative robots that can employ heavier end effectors (currently limited around 5-10 kilograms)
- Smaller robots for the aerospace industry with more flexibility and with longer reach to provide solutions for numerous subassembly applications.

Conventionally Defined Collaborative Robots

These types of robots fall into a number of loose categories: force torque limited, high payload, humanoid, and mobile. However, as in any other emerging field, classifications are vague and boundaries are often blurred. There are a number of manufacturers—both large and small—active in this sector. A representative selection and their capabilities are listed in Table 1. Note that this is not intended to be exhaustive.

Force Torque Limited In these collaborative robots, the maximum force and torque capable of being exerted by each joint is strictly limited to a level, which while uncomfortable, is unlikely to cause significant injury. The levels are set within the current standards and are typically between X and Y direction axes. The robot structure is designed to eliminate potential trap points and have smooth and rounded structural components. However, it must be remembered that as soon as an end effector (e.g., a drill) is attached, the robot is no longer inherently safe nor collaborative. This is the key point that limits the use of collaborative robotics more than any other single factor. An important question we will be discussing further is whether any form of physical contact is tolerable, even if it is safe. The violation and encroachment on personal space is something that can have adverse effects on human operators. The safety systems or design should ensure that the force torque limitation should not be the primary safety mechanism and only come into force as a last line of protection for the operator (Figure 5).

A number of manufacturers are active in this domain and we have included the most common as examples only—it is not the purpose of this report to provide a market survey. The six-axis models manufactured by the Netherlands-based Universal Robots have payload capabilities between 3 and 16 kilograms and are designed for tasks like assembly and light handling. They have been highly successful due to ease of installation and programming. Another example is the LBR iiwa manufactured by KUKA of Germany; this seven-axis robot is available with payload capabilities between 7 and 14 kilograms.

High Payload In this category, the collaborative capability is achieved through the use of external sensors to detect when potential collisions are likely. The sensing capability may either be external to or integral with manipulator. These robots are usually based on existing industrial robots and offer potentially higher payload capacity and a more robust design, but applications must be designed with great care to ensure safety and compliance with standards and regulations. At their simplest, they are programmed to slow down and eventually stop if a human enters their safe working envelope.

FIGURE 5. Twin arm semi-humanoid collaborative robot.

This feature is often available as an add-on (such as KUKA SafeRobot). It could be argued that these types are not really collaborative, but they are mentioned here for completeness. An example of a truly collaborative robot in this category is the Japanese FANUC CR-35iA. This is a standard FANUC manipulator, but the base of the robot is modified and fitted with force sensors to compare actual against expected force on the manipulator, and to stop when an imbalance occurs. This provides a potential payload of up to 35 kilograms, which is significantly greater than the more conventional collaborative robots (Figure 6).

Humanoid These are the classic anthropomorphic robots of science fiction that seek to mimic the physical behavior of human beings. Humanoid robots have largely been confined to laboratory demonstrations and have received—until now—only limited attention for industrial applications. This is due to issues regarding stability, energy consumption, performance, and the like. However, they are being investigated more seriously as the technology begins to improve and present potentially viable solutions. Many of the same constraints and issues identified in this report will also apply to them as they move from the laboratory. A number of robots have been developed that effectively model the human upper body and arms allowing them to manipulate objects in much the same way as humans do. Examples include the Swiss-Swedish ABB YuMi and the Japanese Yaskawa MOTOMAN DA20.

Exoskeleton It is questionable whether robotic human exoskeletons are really collaborative robots; however, they are included here for completeness. They are in many ways the closest physical interaction between robotics and humans as a wearable extension of the operator's body. They provide assistance through the enhancement of physical strength but are not capable of—and indeed probably should

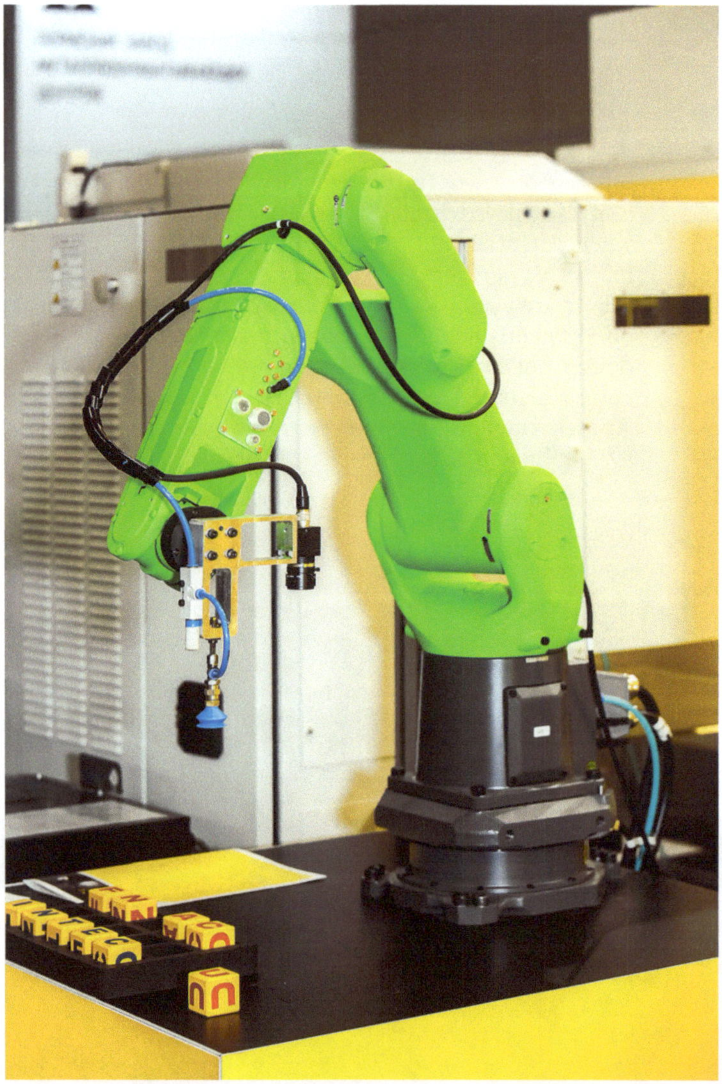

Alexander Tolstykh/Shutterstock.com

FIGURE 6. FANUC collaborative robot.

not be able to—perform independent actions. A number of industrial exoskeleton systems are under development for industrial applications, such as the Italian Comau Exoskeleton, MATE.

Mobile In industrial applications, these are usually hybrid systems with a collaborative arm mounted on a mobile platform. There are two separate safety requirements as the platform must be able to navigate safely while the manipulator has to operate within the same constraints as any other collaborative robot. As mobile systems are being developed with increasing autonomy, they are bringing new requirements for safety standards to cover more dynamic speed and separation safety monitoring.

Collaborative Installations

As we have already asserted, any robot may be operated collaboratively with the necessary control and sensing infrastructure to prevent inadvertent and potentially dangerous contact between the robot and operator. Applications like these typically use external sensors to detect and locate incursions within the working envelope of the robot. A typical example is the SafetyEYE system developed by Pilz, which uses a camera system to monitor a preprogramed area [9]. Another example is the laser-based systems such as those manufactured by SICK [10]. The camera-based systems are primarily three-dimensional, while the laser-based systems are predominantly two-dimensional. New systems based around radio frequency detection are also being introduced, which are more robust and less affected by ambient light and reflectivity. Other approaches include pressure sensitive mats or flooring, but these tend to have less sensitivity and often are limited to demarcating "GO/NO GO" zones [11].

The integration of sensors is key to how the cell can be used; the robot may have software (developed by the original equipment manufacturer) on the controller that controls the speed of the manipulator depending on the distance of the incursion from the manipulator. This provides a dynamic environment so that standards compliance can be maintained—the closer the operator is, the lower the maximum robot speed, thus upholding the effective safe stopping distance. Another approach is to configure a series of zones that can be used to warn an operator when an incursion occurs or to trigger an emergency stop depending on the proximity and speed of the robot within that zone. This provides a significant level of flexibility, but a key requirement is how to inform the operator which areas have which status or approach. Fixed floor markings can be used, but these eliminate the flexibility and capability to manage protection zones dynamically.

In practice, many fully collaborative robot cells will need to incorporate an additional active safety system to provide operator protection from process-based risks such as ejection or rotating tools. A high degree of redundancy is necessary within the safety system to prevent potential unguarded areas occurring due to obstruction of sensors. The operation of the cell must be carefully considered when deciding this as potential sensor obstructions may occur during robot movement or operator actions.

Recommendations

It is becoming clear that the more conventional idea of a "cobot" as a specially designed, human-safe robot is impractical for most applications, particularly in aerospace manufacturing. There may be some traction for this concept in lightweight assembly or packaging applications, but for the majority of processes we wish to integrate robots into either the required loads far exceed the typical 10 to 15 kilogram-limits, or the process itself introduces unacceptable risk. In reality, all robots have the potential to be collaborative and the level of possible collaboration depends on the ability to detect and react to the presence of the human in the same manner as when human operators are working together in close proximity. This requires controllers and manipulators with built-in sensory capabilities or the control capability to react rapidly—in a safe way—to potential unintended contact. In addition to the robot, we need advanced external sensing capabilities that can provide full, incremental, three-dimensional coverage of workspaces without obstructions or dead spaces. The issue is perhaps choosing what type of robots we need (e.g., for airframe assembly). What reach, payload, accuracy, repeatability, or degrees of freedom do we need and how we make them collaborative to optimize their deployment?

Effective Application and Task Division between Humans and Robots

To fully exploit human-robot collaboration, we need to have a clear understanding of how we can divide the proposed task or process between the two elements. Engineers are accustomed to designing the coordination of machines and humans separately, but balancing workflow between an intelligent robot and a human requires new knowledge about how they will interact. Simply allocating the most basic tasks to the robot and the complex ones to the human is not necessarily the best approach. It is also critical to ensure the interaction maintains balanced pace and timing that suits both elements. Also we need to consider carefully whether human-robot collaboration is actually the most appropriate course. We need a clear methodology to first validate if a task is suitable for collaborative working, and second, for the optimal selection and division of tasks between human and robot. In reality there are many reasons why we would want to automate parts of a task:

- Automated processes tend to be repetitive.
- Automated processes tend to be relatively simple (i.e., they do not require process feedback loops and adjustments).
- Automated processes may help with health and safety issues, such as lifting, or to prevent repetitive strain injury.
- Manual things tend to require dexterity or feedback-adjustment control, which people can easily achieve.
- Manual work requires craftsmanship, while automated processes tend to focus on simple nonskilled work. (However, this last statement needs to be treated with care as in many cases it is very dependent on what we define as craftsmanship.)

While these are all valid assertions, automation is—in practice—usually driven by a complex balance between cost and complexity, and the need to build both a viable process and business case. Unless we consider all aspects of the process and its implementation within a facility, it is unlikely to be successful. It is also important that we consider the human operator's relationship to and perception of ownership of tasks.

Another obstacle to evaluations of potential applications and task division is that we are still not able to adequately evaluate human functions. It is relatively straightforward to assess and allocate physical capabilities and this remains relevant for ergonomic safety [12]. Additionally, a decompositional systems engineering approach remains useful for identifying the interactions and requirements of constituent parts. However, the greater cognitive demands involved in more cooperative human-robot task-sharing require more complex analysis in order to assess and assign higher-order human cognitive processes appropriately. Recent attempts to develop systematic techniques based on task classification and abstraction show signs of potential but have yet to successfully address the sophistication and variability of human cognitive skills [13, 14]. Put simply, we are not yet capable of systematically classifying and quantifying human cognitive functions across different tasks and robot applications.

Recommendations

The key to achieving the best possible task division between humans and robots is to understand the task required. However, this is not just from a physical/motion perspective but also from the point of view of cognitive load. We need to be able to assess the cognitive demand placed on the operator, not only by the task, but also by the collaboration itself. Further research is needed to systematically classify the physical interactions required and also the human cognitive functions involved. Once we have a clear representation of the task, it will then become easier to map elements to the human and robots. However, it will still not be a trivial effort. Further work is required to ensure that this part of the process is optimized as it is central to effective and long-term acceptance. If an application is suboptimal then it will eventually be rejected by operators resulting in resistance and lowered productivity.

Building an Effective and Credible Safety Case for the Implementation of Collaborative Robots

A misguided and simplistic attraction of collaborative robotics is the misconception that they are cheaper and easier to install than conventional robots, as they do not need additional external safety systems. This, in turn, results in a potentially smaller footprint and allows closer proximity to human workers. This is unfortunately only true in the case of the simplest applications, such as packing. As we have highlighted, it is often the process that is more dangerous than the robot. Therefore, the same level of attention must be paid to the design of the safety systems, and in many cases, it is more complex due to dynamic nature of such cells. To satisfy regulatory requirements and generate organizational and worker confidence, a comprehensive risk assessment needs to be developed to demonstrate that collaborative installations can be operated safely. This is often problematic, and acceptability can vary considerably depending on many different factors such as company culture and risk acceptability. It has been suggested during the preparation of this SAE EDGE Research Report that the mind-set toward traditional health and safety assessments—in many cases—needs to be much more objective and aligned with the technology and standards in place. It is sometimes easier for the responsible person to simply say he is not happy with the application even though it may meet all safety standards and requirements. Within the industry, we have noted many common requirements emerging for the implementation of collaborative robots in an acceptable way (Figure 7):

- Standardized way of measuring the safety of a collaborative robot.
- Divide the risk assessment of the collaborative robot operations and the process risk assessment.
- Develop a more dynamic approach to risk assessments as the process and its constituents changes.

Additionally, the feasibility of industrial human-robot collaboration was first defined in the ISO 10218-2:2011

FIGURE 7. **Conventional physically protected robot installation.**

Vydrin/Shutterstock.com

standard, which specifies four possible shared workspace conditions to advance integration:

- **Safety-rated Monitored Stop:** where the robot workspace is stopped via sensor/vision monitoring systems when human incursion into the predefined workspace is detected
- **Speed and Separation Monitoring**: where sections of the workspace are monitored by sensor/vision systems to allow reduced activity or a full stop depending on the detected human location
- **Hand Guiding:** where the operator activates an enabling switch to initiate a mode for freely moving the manipulator while the robot is still active but without the robot being able to exert additional force or movement
- **Power and Force Limits**: where the system is designed with inherent monitoring, joint compliance, and force torque limitation features to restrict collision impact potential

Subsequently, the ISO Technical Specification ISO/TS 15066:2016 provides greater supplementary detail on how to apply these general requirements. For example, where ISO 10218-2 had merely stipulated that human-robot contact "should not create an injury," TS 15066 included specific threshold pressure and force parameters for "quasi-static and transient contact" across different parts of the body, to inform the design and risk assessment of energy transfer and speed limits. However, to date, no safety standards attempt to adequately address aspects of psychological safety and comfort and neither does the current best practice in risk assessment. Available guidance on psychological safety is generic and does not directly address working with robots, collaborative or otherwise [15].

Traditional risk assessment methodologies are mechanistic in their approach as they work by identifying a risk and then mandate how the operator is to be protected, usually by physical exclusion or separation. They tend to be simple and based on checklists and risk matrices [16]. In more dynamic collaborative environments, safety assessment—in the first instance—requires an understanding of how to make collaborative robots safe around humans through the concepts of force, contact pressure, and transfer of momentum for all predictable contact events. Sometimes contact is desirable, such as a handoff of a part, so risk assessment must predict contact that are both planned and due to human tendencies and reasonably foreseeable misuse (i.e., operators do not always follow instructions). However, as dynamics in the environment increase, a more systemic view is needed, which considers the performance of the system as a whole—including interactions between human, technical, psychological, and environmental factors. Emerging system-based models for complex sociotechnical systems include the Functional Resonance Analysis Method and Systems-Theoretic Accident Model and Processes [17]. These models tend to consider systems as a whole and take in to account the interaction of the operator, environment, and robot. Furthermore, they can also account for interactions between systems and the organization that they are in, as culture and attitudes toward risk and mitigation can vary considerably, even in very similar domains [18]. Indeed, attitudes to safety can be influenced as much by culture and employee experience as by standards and protocols. A more task-oriented approach has also been proposed, which uses task analysis and decomposition to complete the risk assessment process before the cell is realized [19].

Standard risk assessment techniques are needed to ensure consistency of best practice and principles. The more we design and assess collaborative robot systems, the more we can identify commonalities between them and develop new standard practices and principles. To improve aerospace sector confidence in collaborative systems, experiential knowledge needs to be developed quickly.

Recommendations

The complexity of the risk assessment and associated challenges can be a deterrent for collaborative applications. The addition of human behavior and all of its potential variability and unpredictability can be a significant obstacle to confidence. What is needed is a practical methodology for risk assessing collaborative applications that is proactive in its management of risk and that can manage dynamically changing systems. It must be able to satisfy regulatory bodies and be robust enough to stand up in a court of law if necessary. Standard risk assessment techniques that include a balanced evaluation of robot and human dynamics are needed to ensure consistency of best practice and principles.

Understanding the Human Aspects of Human-Robot Collaboration

The fundamental aim of collaborative robotics is to combine and utilize the unique strengths of both the human and the robot. In almost all cases, the human role is to supply cognitive skills and flexibility (i.e., decision-making, monitoring, remedial interventions, etc.). The problem is, there is still no unified understanding of how human cognition is impacted by robot collaboration, nor common method for evaluating human requirements for different systems. Organizations need confidence to deploy collaborative robotics; they need to know a new system on the shop floor will not bring any negative impacts to their workforce. Primarily, they need knowledge and tools to optimize trust and acceptance when new systems are introduced and to optimize collaborative task performance after systems are operational (Figure 8).

Trust and Acceptance

Trust has long been recognized as a crucial element in human-robot interaction within the service and assistive robotics domain—but it has not been addressed until recently in industry due to the strict separation of robots and people in traditional manufacturing systems. However, it has been shown that trust now plays a critical role in the success of new industrial automation [20]. We have long known that levels of trust can be affected by a number of factors, including system design characteristics like robot size and appearance, speed and velocity, human-like movements, and proximity to the person [21, 22, 23, 24, 25]. In new encounters with industrial robots, trust is a function of perceptions related to the system's motion and speed, co-operation, and reliability [26, 27]. Although organizations know that workers naturally develop confidence in using equipment over time, they need to promote trust at an early stage to prevent costs caused by slow acceptance and uptake. Engaging operators in the implementation of a new collaborative system gives them the opportunity to touch and feel their new robotic co-worker and contribute to decisions. Humans can accept the need for new technology but not necessarily trust it, and their perceptions of its usefulness and ease of use will influence whether they actually use it [28]. Thus, personal involvement and investment in the change process can raise awareness regarding a new system's utility and value, while eliminating mistrust and fear over losing personal skills or being replaced by a robot.

A common industrial concern is that older operators will not collaborate with robots because they are believed

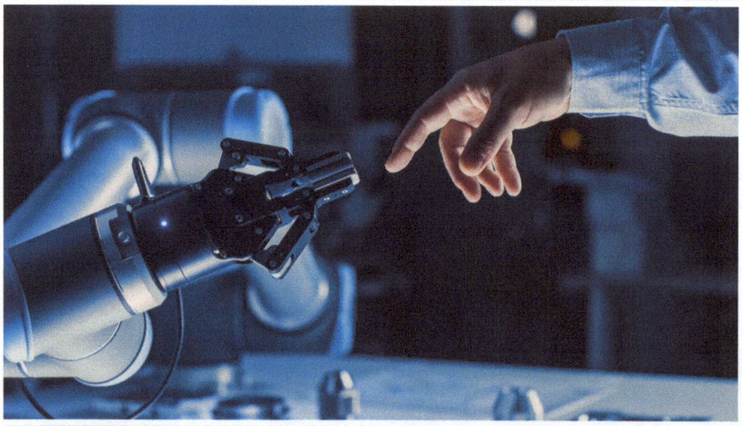

Gorodenkoff/Shutterstock.com

FIGURE 8. How do human interact?

to be more resistant to change and new technologies than younger people. Although older workers can prefer familiar ways of working and be less enthusiastic about new training activities, age is not significantly related to negative attitudes toward organizational changes and innovation [29]. Moreover, various research initiatives are now developing adaptive and personalized collaborative robotics to accommodate individual operator requirements so future systems will address preferences of the aging workforce along with those of other groups and individuals [30].

Collaborative Task Performance

As evident by their lack of inclusion in safety standards and risk assessment methods, "softer" human factors have traditionally been considered far less important than physical/ergonomic impacts in the industrial context. Nonetheless—and organizations often see this first-hand—psychosocial conditions and responses can be just as influential to both performance and safety outcomes.

Manufacturing system design understandably prioritizes product and process requirements first. Manual work is typically designed later, to fit in with whatever hard system components and technologies have been selected for the production. For true human-robot collaboration, this traditional "fitting the person to the job" approach is not enough, as successful performance relies on coordination and synchronicity of both human and system behaviors. This requires a shared representation and understanding of the problem state, and activities of other agents, to support coordinated team problem solving. Effective task division requires comprehensive understanding of those behaviors and how system characteristics influence task performance. In the same way that system characteristics influence human trust at the start of a collaborative relationship, they can also cause various cognitive/affective responses that impact operators' task performance. For example, proximity can increase anxiety, speed can increase mental workload, and robot reliability can affect confidence [31, 32, 33].

A vast amount of research has explored human-robot interaction and collaboration, but studies have investigated very different aspects and applications, which prevent comparison of results. Moreover, few studies have applied appropriate research methodology for investigating key human responses and psychological safety [34]. A range of reliable human factors methods are useful for exploring tasks and assessing impacts on cognitive/affective processes, such as mental workload, situation awareness, and so forth, but there is no prescribed standard guidance or practical tools.

The first formally published standard to deal with ethical issues of robot design and application, BS 8611:2016, and a forthcoming ISO Technical Report on the "Ergonomics of human-system interaction for Robotic, intelligent and autonomous systems," ISO/TR 9241-810, both address wider human/social implications of collaborative robotics. These documents are an important step in dealing with wider issues, but neither will specifically guide the design of industrial applications.

Recommendations

Trust and acceptance are critical to the success of collaborative robotics. Acceptance of a new system is built on trust, and both are a function of the system's characteristics and conditions—which will also go on to affect human responses and behaviors that determine operational performance. Therefore, it is vital that we develop a contemporary and comprehensive understanding of the impacts of system characteristics and conditions. Industry needs clear and reliable guidance and practical tools to optimize the design and implementation of collaborative systems and safety standards that will safeguard psychological and well-being impacts.

Summary

Within this SAE EDGE Research Report, we have attempted to provide an overview of the current state of implementation of human-robot collaboration within aerospace manufacturing and to seek out the key issues preventing wider uptake and implementation. We have provided a brief overview of some current applications to help set the scene but this is not intended to be a state-of-the-art report. Nor is it intended to be a "how to" guide. Rather its aim is to understand what needs to be done to drive the subject forward so that its true potential is realized. This work has been not only based upon the literature, but more importantly by listening to experts from the industry and robot and ancillary manufacturers. They have shared the thoughts and attitudes they have encountered within the sector and many of their own frustrations.

It is apparent that here are many differing and often conflicting definitions of what human-robot collaboration is and how it is implemented. What is most striking is how broad the definitions are: simple tasks such as fixture loading can be considered alongside complex multiagent interactions. A common taxonomy is needed to allow tasks to be defined. Also, the ubiquitous term "cobot" is misleading and in many ways limits the application of human-robot collaboration by engendering preconceptions of what collaborative robotics is. Another, important consideration is not just the physical elements of the task but the cognitive ones that need to be considered. Furthermore, we need to understand the disruptive effects that can occur and the increased cognitive loading caused by the collaborative environment. Risk assessment and management was another obstacle, leading to the need to construct a robust risk assessment that would support a more dynamic task interaction and environment, while still fully satisfy the regulatory authorities. And finally and perhaps most importantly, how do we account for the human? A lot of applications tend to focus on the robot rather than on the human to the detriment of the overall application. We refer to human/robot collaboration and should therefore have the needs and sensitivity of the human operator at the forefront.

Building on this collective knowledge, we have made a number of recommendations for the way forward to help inform future dialogue and research. We know that work is already in progress in many of these areas, particularly within the standards community and we look forward to the outcome of this. There are also areas that we have not covered, for example, the work in this SAE EDGE Research Report has concentrated on airframes but many of the concepts and identified issues are equally applicable to engine assembly and indeed many other industry sectors.

SAE EDGE™ Research Reports

SAE EDGE™ Research Reports, like the present report on "Unsettled Issues on Human-Robot Collaboration and Automation in Aerospace Manufacturing," are intended to push further out into still unsettled areas of technology of interest to the mobility industry. SAE launches these reports before attempting to form a joint working group, let alone a cooperative research program or a standards committee.

SAE EDGE™ reports are intended to be quick, concise overviews of major unsettled areas where vital new technologies are emerging. An unsettled area is characterized more by confusion and controversy than established order. Early practitioners must confront an absence of agreement. Their challenge is often not to seize the high ground but to find common ground. These scouting reports from the frontiers of investigation are intended merely to begin the process of sorting through critical issues, contributing to a better understanding of key problems, and providing helpful suggestions about possible next steps and avenues of investigation.

SAE EDGE™ Research Reports, therefore, are fundamentally distinct from the more formal working groups approach and far removed from the more mature research program and standards development process.

Next Steps for Human-Robot Collaboration and Automation in Aerospace Manufacturing

This publication should be considered only as a first step toward clarifying the issues around human-robot collaboration in aerospace manufacturing. The intention behind this and other SAE EDGE™ Research Reports is to start a dialogue among interested parties on important industry-wide topics that require further attention. The expectation is that these explorations of unsettled areas of technology will lead to the formation of working groups and, ultimately, committees that can address and resolve the issues they raise, producing a framework for developing a common vocabulary of definitions, best practices, protocols, and standards needed to support continued progress toward safer and more innovative products.

The experts' collaboration that gave rise to this publication demonstrated a great willingness on the part of the industry to define the terminology, procedures, and eventually the standards needed to enable human-robot collaborative technology to move ahead as quickly and efficiently as possible.

This SAE EDGE™ Research Report on "Unsettled Issues on Human-robot Collaboration and Automation in Aerospace Manufacturing" identifies the following key topics for further pursuit, both through continued informal discussions among industry practitioners and through more formal working groups:

- Develop a unified taxonomy that can be used to define modes and degrees of human-robot collaboration across different domains.
- Develop controllers and manipulators with built-in sensory capabilities or the control capability to react rapidly in a safe way to potential unintended or anticipated contact.
- Standardize risk assessment techniques to ensure consistency of best practice and principles.
- Work to systematically classify not only the physical interactions required but also the human cognitive functions involved and the impact of collaborative working in itself.

Recommendations

The overall recommendations of this SAE EDGE™ Research Report can be summarized as follows:

1. There is a clear need for a unified and internationally agreed taxonomy for human-robot collaboration and the degrees of interaction.
2. The international standards need to review and revise on a much more frequent basis in order to keep pace with the rapidly developing technologies. There is a significant risk that the field may be held back by outdated standards that do not reflect the actuality of the technology.
3. More flexible and active monitoring technologies need to be developed and deployed, otherwise the superior lifting and manipulation capabilities of robots cannot be fully exploited. Collaborative robotics will be confined to lightweight, low force tasks.
4. The division and allocation of tasks between human and robot needs to be better understood to maximize effective and successful interaction. Ideally the

interaction should be able to use the physical and mechanical advantages of the robot without reducing the skill and satisfaction of the human.

5. More emphasis must be placed upon the interaction with the human and their reaction and emotional response. Until now, the majority of research work has focused on the technology and the physical safety of the human.

Acknowledgments

Recognition should go first to all the participants, many of who also provided feedback on the draft version of this publication. Without their input and initiative, this SAE EDGE™ Research Report would not have been possible.

Andrew Armstrong, Fanuc Robotics
Marco Chacin, PhD, Airbus
Elena Dominguez, Pilz
Joshua Downs, Boeing
Leo Muijs, GKN Fokker
Andrew Portsmore, GKN Aerospace

The authors of this document together with the SAE Team responsible for its creation join in expressing our deepest appreciation to all the individuals mentioned above.

Professor Philip Webb, PhD
Cranfield University

Sarah Fletcher, PhD
Cranfield University

References

1. Barattini, P., Vicentini, F., Virk, G.S., and Haidegger, T., editors, *Human-Robot Interaction: Safety, Standardization, and Benchmarking* (Boca Raton: CRC Press, 2019). ISBN:1351819631.
2. EFFRA, "Validation of Advanced, Collaborative Robotics for Industrial Applications," https://portal.effra.eu/project/1038, accessed Oct. 26, 2020.
3. Muijs, L., and Snijders, M., "Collaborative Robot Applications at GKN Aerospace's Fokker Business," SAE Technical Paper 2017-01-2091, 2017, https://doi.org/10.4271/2017-01-2091.
4. Boeing, "Mechanic and Machine: Boeing's Advanced Manufacturing Improves 777 Assembly," Feb. 2017, https://www.boeing.com/features/2017/02/faub-777-assembly-02-17.page, accessed Oct. 26, 2020.
5. Gates, D., "Boeing Abandons Its Failed Fuselage Robots on the 777X, Handing the Job Back to Machinists," *Seatle Times*, Nov. 13, 2019, https://www.seattletimes.com/business/boeing-aerospace/boeing-abandons-its-failed-fuselage-robots-on-the-777x-handing-the-job-back-to-machinists/.
6. Kheddar, A., Roa, M.A., Wieber, P.-B., Chaumette, F. et al., "Humanoid Robots in Aircraft Manufacturing—The Airbus Use Cases," *IEEE Robotics and Automation Magazine* 26(4):30-45, Dec. 2019.
7. Haddadin, S., and Croft, E., "Physical Human-Robot Interaction," *Springer Handbook of Robotics* (2016), 1835-1874, https://doi.org/10.1007/978-3-319-32552-1_69.
8. Yanco, H., and Drury, J., "Classifying Human-Robot Interaction: An Updated Taxonomy," *IEEE International Conference on Systems, Man and Cybernetics* 3:2841-2846, 2004, https://doi.org/10.1109/ICSMC.2004.1400763.
9. Pilz, "Safe Camera System SafetyEYE," https://www.pilz.com/en-GB/eshop/00106002207042/SafetyEYE-Safe-camera-system, accessed Oct. 26, 2020.
10. SICK, "Safety Laser Scanners," https://www.sick.com/gb/en/opto-electronic-protective-devices/safety-laser-scanners/c/g187225, accessed Oct. 26, 2020.
11. Rockwell, "Safety Mats," https://ab.rockwellautomation.com/Sensors-Switches/Operator-Safety/MatGuard-Safety-Mats#applications, accessed Oct. 26, 2020.
12. El Makrini, I., Merckaert, K., De Winter, J., Lefeber, D. et al., "Task Allocation for Improved Ergonomics in Human-Robot Collaborative Assembly," *Interaction Studies* 20(1):102-133, 2019.
13. Roncone, A., Mangin, O., and Scassellati, B., "Transparent Role Assignment and Task Allocation in Human Robot Collaboration," in *2017 IEEE International Conference on Robotics and Automation (ICRA)*, Singapore, 1014-1021, IEEE, 2017.
14. Micheler, S., Goh, Y., and Lohse, N., "A Transformation of Human Operation Approach to Inform System Design for Automation," *Journal of Intelligent Manufacturing* 1-20, https://link.springer.com/article/10.1007/s10845-020-01568-z.
15. Canadian Standards Association, *Psychological Health and Safety in the Workplace—Prevention, Promotion, and Guidance to Staged Implementation* (Quebec, ON: Bureau de normalisation du Québec, 2013).
16. Caputo, A.C., Pelagagge, P.M., and Salini, P., "AHP-Based Methodology for Selecting Safety Devices of Industrial Machinery," *Safety Science* 53:202-218, 2013.
17. Leveson, N., "A New Accident Model for Engineering Safer Systems," *Safety Science* 42(4):237-270, 2014.
18. Kawar, "Adoption of Robotic Innovation: Assessing the Safety Risk of Large Scale Collaborative Automation and Measuring its Acceptance," Ph.D. thesis, Cranfield University, 2017.

19. Marvel, J., Falco, J., and Marsti, I., "Characterizing Task-Based Human-Robot Collaboration Safety in Manufacturing," *IEEE Transactions on Systems, Man and Cybernetics: Systems* 45(2):260-275, Feb. 2015.
20. Charalambous, G., Fletcher, S., and Webb, P., "Identifying the Key Organisational Human Factors for Introducing Human-Robot Collaboration in Industry: An Exploratory Study," *The International Journal of Advanced Manufacturing Technology* 81(9-12):2143-2155, 2015.
21. Hancock, P.A., Billings, D.R., Schaefer, K.E., Chen, J.Y.C. et al., "A Meta-Analysis of Factors Affecting Trust in Human-Robot Interaction," *Human Factors* 53(5):517-527, 2011.
22. Butler, J.T., and Agah, A., "Psychological Effects of Behavior Patterns of a Mobile Personal Robot," *Autonomous Robots* 10(2):185-202, 2001.
23. Nonaka, S., Inoue, K., Arai, T., and Mae, Y., "Evaluation of Human Sense of Security for Coexisting Robots Using Virtual Reality. 1st Report: Evaluation of Pick and Place Motion of Humanoid Robots," in *IEEE International Conference on Robotics and Automation, 2004. Proceedings. ICRA'04*, New Orleans, LA, 2004, vol. 3, 2770-2775, IEEE, 2004.
24. Zanchettin, A.M., Bascetta, L., and Rocco, P., "Acceptability of Robotic Manipulators in Shared Working Environments through Human-Like Redundancy Resolution," *Applied Ergonomics* 44(6):982-989, 2013.
25. Bartneck, C., Kulić, D., Croft, E., and Zoghbi, S., "Measurement Instruments for the Anthropomorphism, Animacy, Likeability, Perceived Intelligence, and Perceived Safety of Robots," *International Journal of Social Robotics* 1(1):71-81, 2009.
26. Charalambous, G., Fletcher, S., and Webb, P., "The Development of a Scale to Evaluate Trust in Industrial Human-Robot Collaboration," *International Journal of Social Robotics* 8(2):193-209, 2016.
27. Story, M., Jaksic, C., Fletcher, S., Webb, P. et al., "Evaluating the Use of Human Aware Navigation in Industrial Robot Arms," in Tokhi, M.O., Ferreira, M.I.A., Govindarajulu, N.S., Silva, M. et al. (Eds.), *Artificial Intelligence, Robots and Ethics - Proceedings of the Fourth International Conference on Robot Ethics and Standards (ICRES 2019)*, 2019, CLAWAR Association Ltd., 79-86.
28. Maurtua, I., Ibarguren, A., Kildal, J., Susperregi, L. et al., "Human-Robot Collaboration in Industrial Applications: Safety, Interaction and Trust," *International Journal of Advanced Robotic Systems* 14(4):1729881417716010, 2017, https://doi.org/10.1177/1729881417716010.
29. Ng, T.W.H., and Feldman, D.C., "Evaluating Six Common Stereotypes about Older Workers with Meta-Analytical Data," *Personnel Psychology* 65(4):821-858, 2012.
30. European Union Ace Cluster Factories, "Human-Centred Factories from Theory to Industrial Practice. Lessons Learned and Recommendations," http://ace-factories.eu/wp-content/uploads/ACE-Factories-White-Paper.pdf, accessed Oct. 26, 2020.
31. Arai, T., Kato, R., and Fujita, M., "Assessment of Operator Stress Induced by Robot Collaboration in Assembly," *CIRP Annals* 59(1):5-8, 2010.
32. Koppenborg, M., Nickel, P., Naber, B., Lungfiel, A. et al., "Effects of Movement Speed and Predictability in Human-Robot Collaboration," *Human Factors and Ergonomics in Manufacturing & Service Industries* 27(4):197-209, 2017.
33. Wright, J.L., Chen, J.Y.C., and Lakhmani, S.G., "Agent Transparency and Reliability in Human-Robot Interaction: The Influence on User Confidence and Perceived Reliability," *IEEE Transactions on Human-Machine Systems* 50(3):254-263, 2019.
34. Lasota, P.A., Fong, T., and Shah, J.A., "A Survey of Methods for Safe Human-Robot Interaction," *Foundations and Trends® in Robotics* 5(4):261-349, 2017.

Contact Information

EDGEresearch@sae.org.

www.ingramcontent.com/pod-product-compliance
Lightning Source LLC
LaVergne TN
LVHW070941210926
848482LV00003B/277